Double Talking Helix Blues

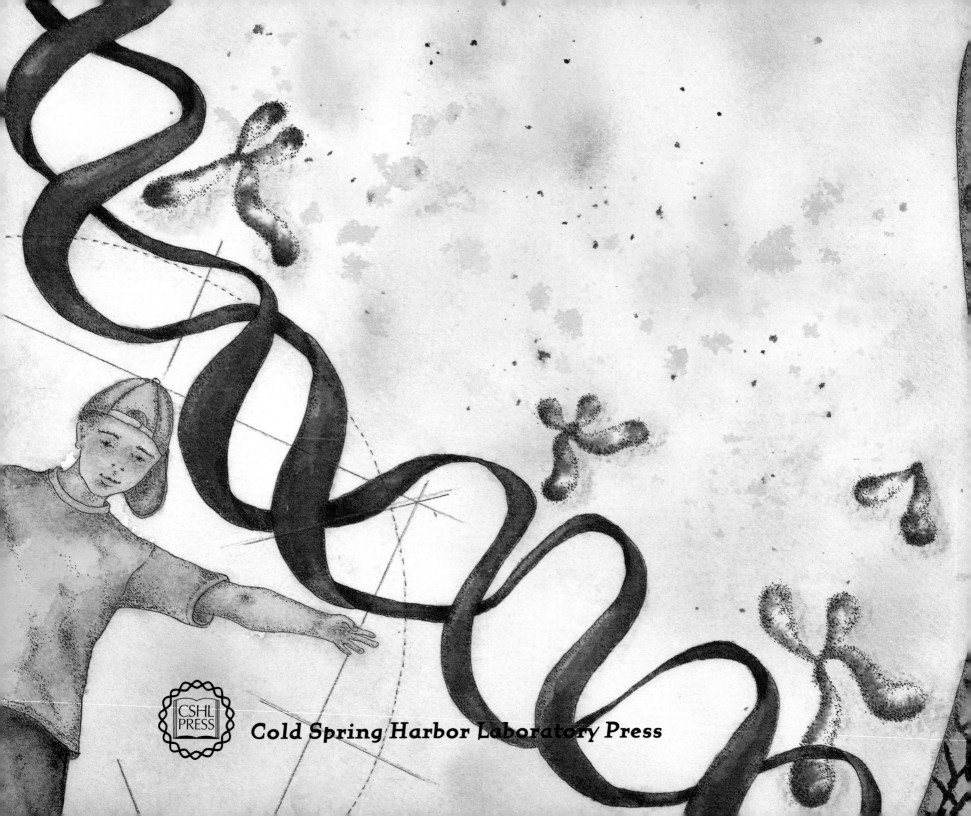

Cold Spring Harbor Laboratory Press

Double Talking Helix Blues

by Joel Herskowitz
illustrated by Judy Cuddihy
additional text by Ira Herskowitz

DOUBLE TALKING HELIX BLUES

Text © 1993 by Joel Herskowitz and Ira Herskowitz

Art © 1993 by Judith G. Cuddihy

Compilation © 1993 by Cold Spring Harbor Laboratory Press

Printed in the United States of America

Library of Congress Cataloging-in-Publication Data

Herskowitz, Joel.
    Double talking helix blues / Joel Herskowitz : illustrated by Judy
Cuddihy.
       p.   cm.
    Summary: Illustrated verses explore the role of DNA in shaping the
formation of new life.
    ISBN 0-87969-431-9
    1. Human reproducton—Juvenile poetry.   2. DNA—Juvenile poetry.
3. Children's poetry, American.   [1. DNA—Poetry.   2. Genetics—
Poetry.   3. American poetry.]   I. Cuddihy, Judy, ill.   II. Title.
PS3558.E792D68   1993
574.87'322—dc20   93-36775
                                       CIP
                                       AC

ISBN 0-87969-431-9

All Cold Spring Laboratory Press publications may be ordered directly from Cold Spring Harbor Laboratory Press, 10 Skyline Drive, Plainview, New York 11803. Phone: 1-800-843-4388 (Continental U.S.
and Canada). All other locations: (516) 349-1930. FAX: (516) 349-1946.

Introduction to the Reader

Did you ever wonder how it is that human beings always have human babies, not carrots or something else? This is a question—even a mystery—that has fascinated people for thousands of years.

The answer to this question comes from looking deep inside cells. Here are located special plans for how each organism will grow. Human cells contain instructions for producing a human. Rabbit cells contain instructions for producing a rabbit.

What are these amazing instructions? How are they copied? The answers to these questions come from understanding the structure of a molecule called DNA. In this book and through the song, "The DOUBLE Talking HELIX Blues," you will learn about DNA—how it is copied and how it works. Here's a clue: DNA has **two** chains. Read on and listen to the tape to see how this number provided one of the keys to unlocking the secrets of DNA.

There may be some words and ideas in this book that you don't know yet. These words are explained in the "Double Helix Guide" at the end of the book. If you want to find out still more about DNA, cells, and molecules, a list of books at the end will help you find more information on these topics.

We hope you will enjoy this introduction to DNA. So, turn on your reading light, plug in your tape player, and have fun!

Well, I once saw in an Addams cartoon
a Martian sittin' in a waiting room.
It was late at night in the maternity ward,
a nurse appeared as she opened the door—
     "Congratulations," she said "it's a baby."

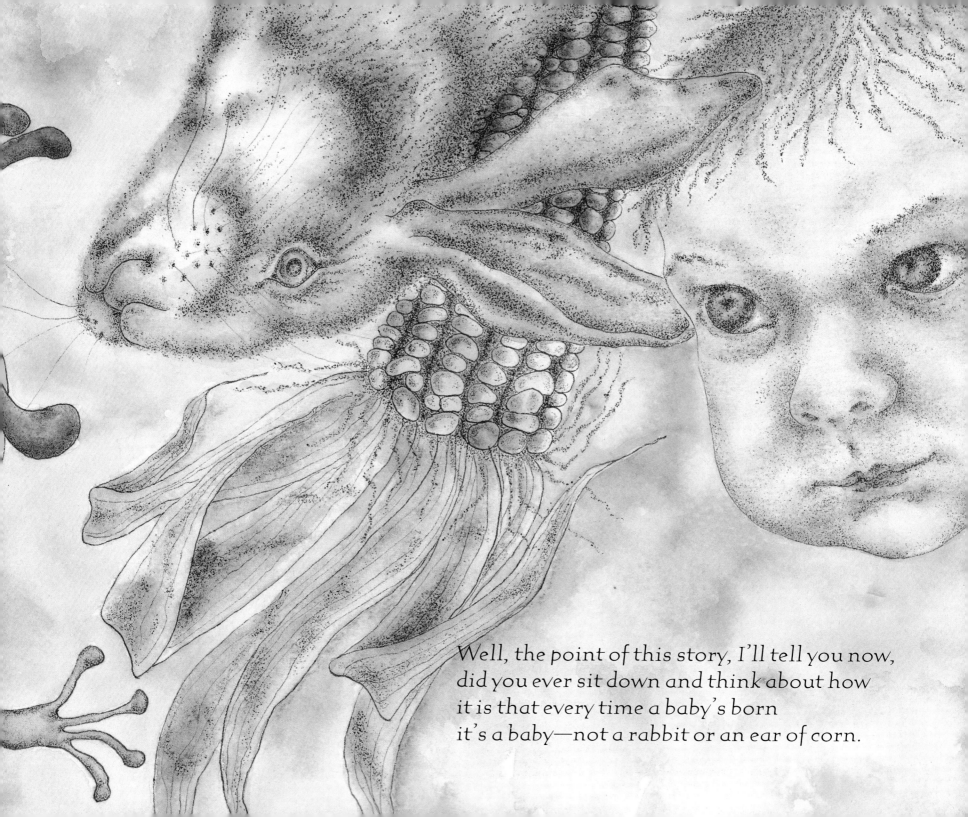

Well, the point of this story, I'll tell you now,
did you ever sit down and think about how
it is that every time a baby's born
it's a baby—not a rabbit or an ear of corn.

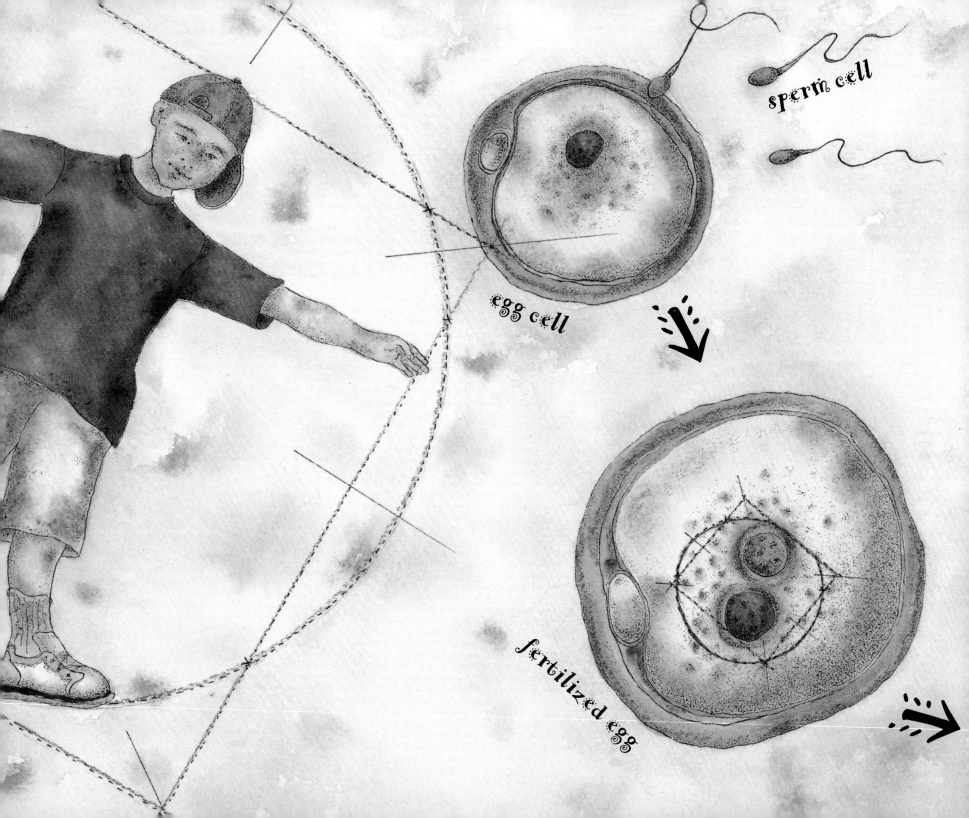

sperm cell

egg cell

fertilized egg

Well, it just so happens that inside everyone
are tiny plans that tell how the job's to be done:
They're worth more to you than the family jewels.
They're stored in the form of molecules.
    Like everything else I guess.
    Only different, and kind of special.

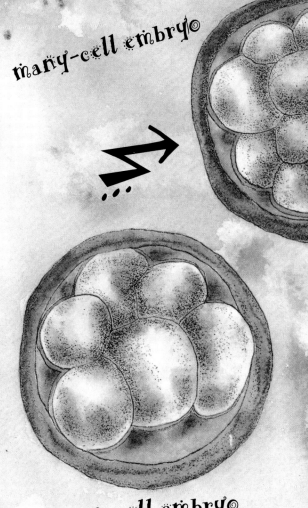

many-cell embryo

8-cell embryo

4-cell embryo

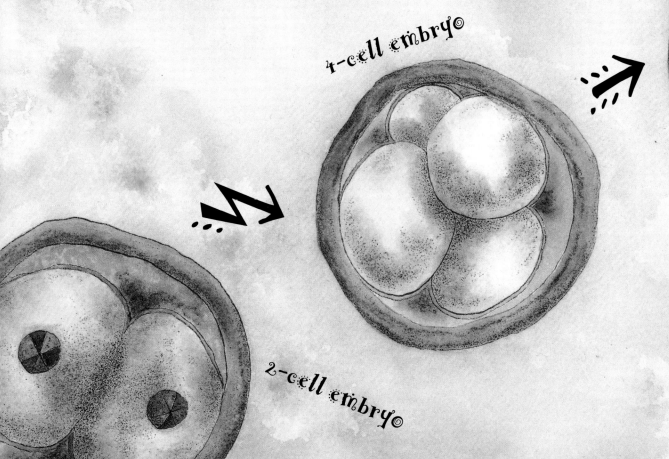

2-cell embryo

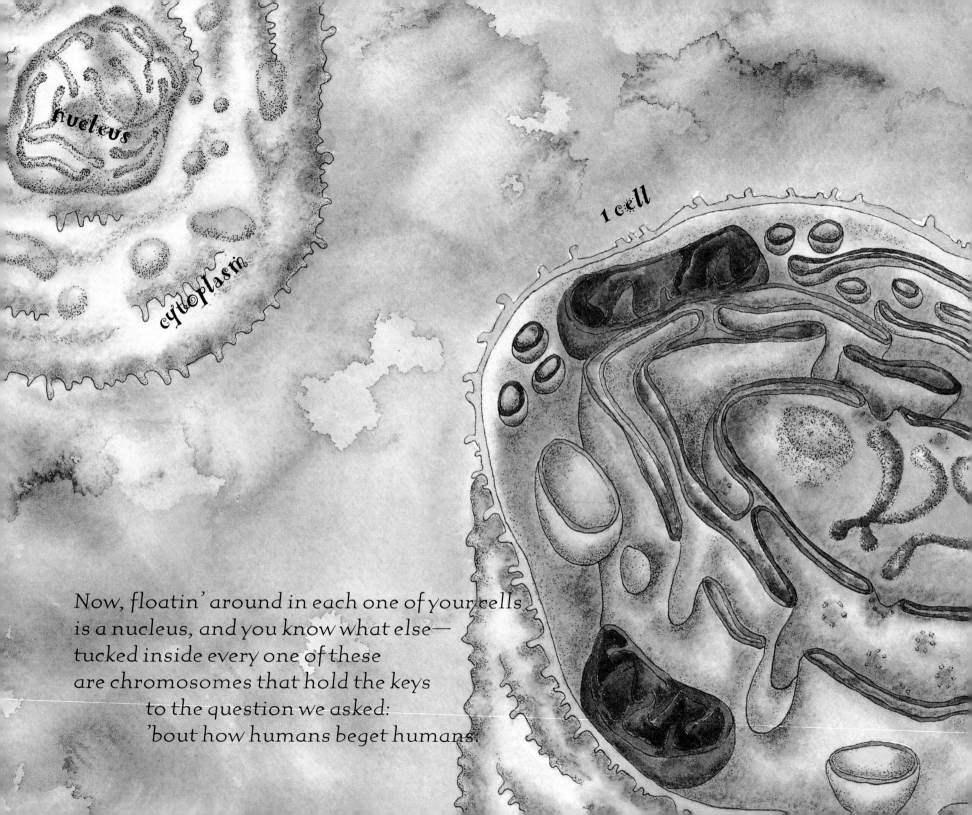

nucleus

cytoplasm

1 cell

Now, floatin' around in each one of your cells
is a nucleus, and you know what else—
tucked inside every one of these
are chromosomes that hold the keys
to the question we asked:
'bout how humans beget humans.

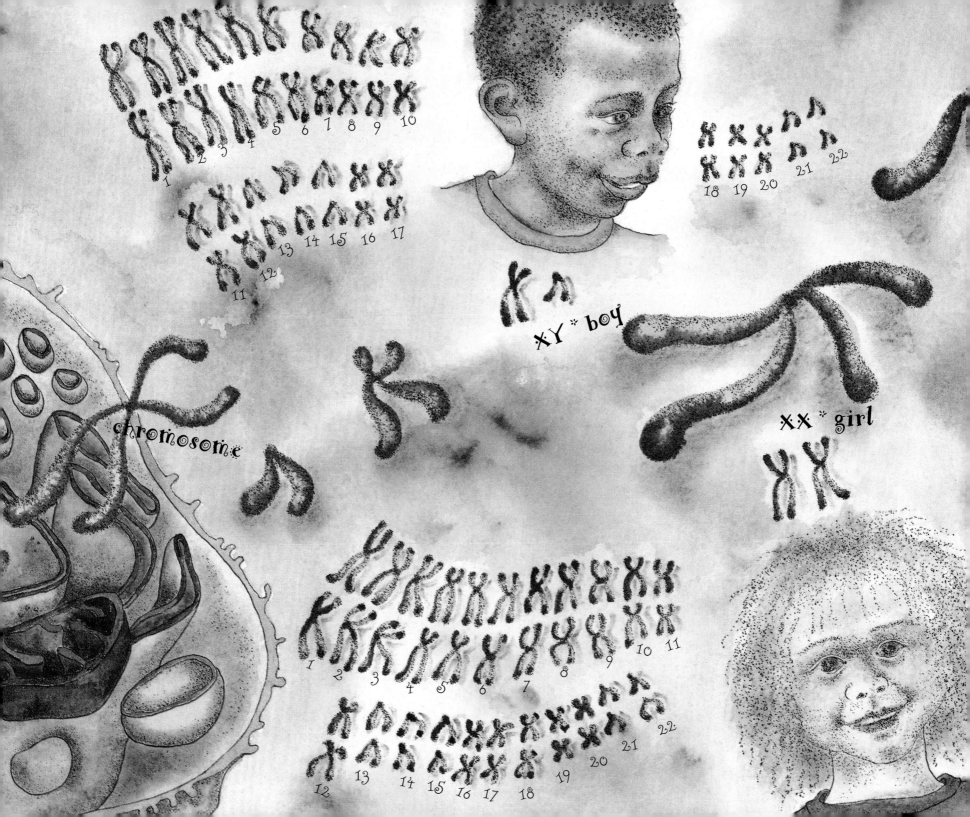

chromosome

XY * boy

XX * girl

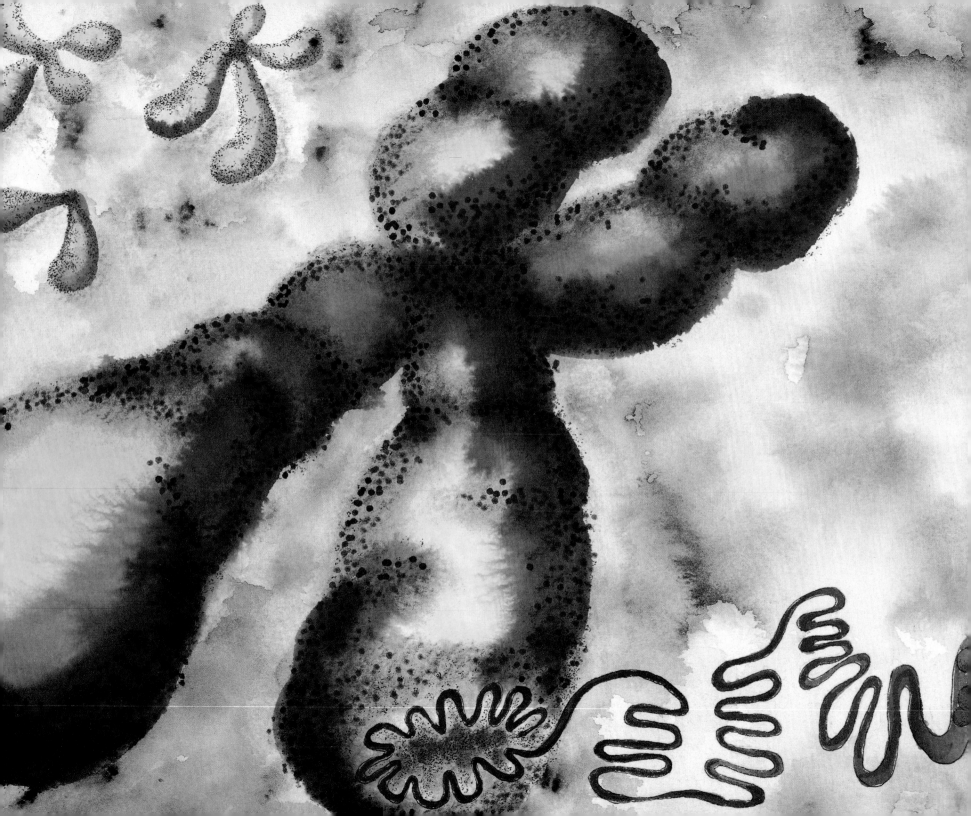

Now, these threadlike chromosomes contain
a groovy little substance that we're gonna name;
it's a macromolecule, as they say.
It's mighty fine stuff—called DNA.
        That's "deoxyribonucleic acid"
        For those of you out there with expanding minds.

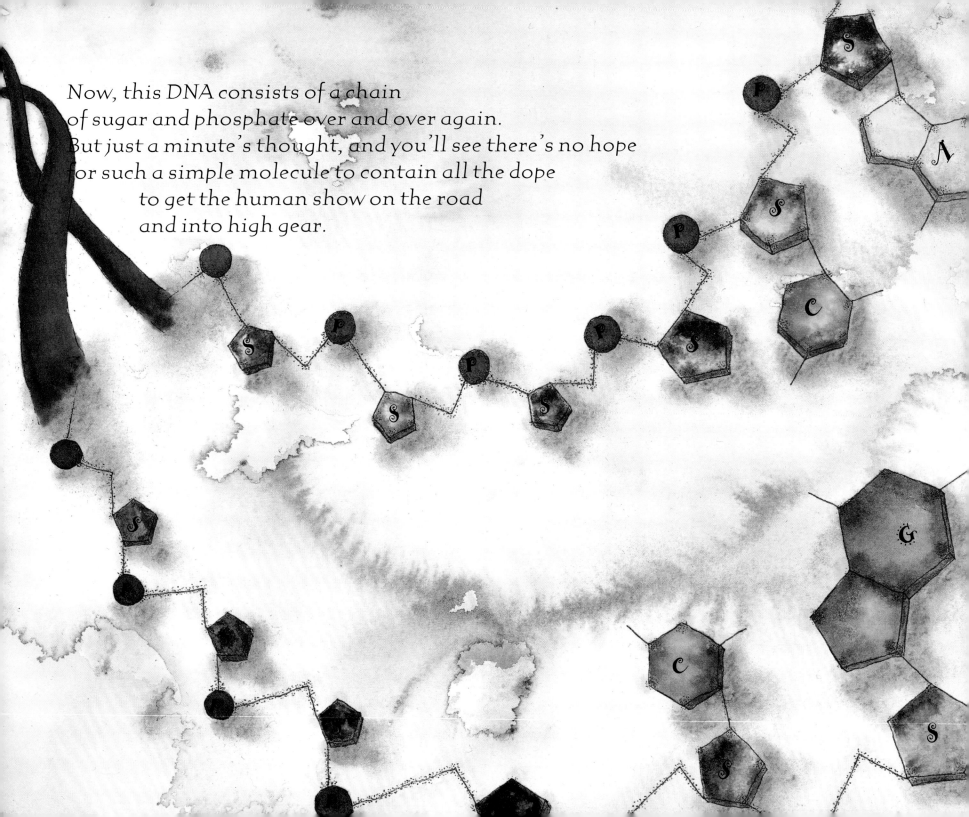

Now, this DNA consists of a chain
of sugar and phosphate over and over again.
But just a minute's thought, and you'll see there's no hope
for such a simple molecule to contain all the dope
to get the human show on the road
and into high gear.

Well, each of the sugars in the backbone
has a ring-shaped base 'tached to it all its own.
There's only four in a human being:
Adenine, guanine, cytosine, and thymine.
    They make up an alphabet of four letters—A, G, C, T.
    Wouldn't want to write a novel with four letters.
    Think I'll write a human being instead!

TTCTGCTATGACCATGACACGATTTAAATCTTTTCAAATGTTTTAGGAGTATTAATCAACATTGTATTCAGCTCTTAAGGCACTAGTGCCT

ATTTAAATATTTTTAAAATATTATTTATTTAACTATTTATAAAACAACTTATTTTTGTTGTCATATTATGTCATGTGCACCTTTGCACA

TGTATTTGGTAAATTTATTTTGTGTTGTTCATTGAACTTTTGCTATGGAACTTTTGTACTTGTTTATTCTTTAAAATGAAATTCCAAGCCTA

GGTACACTTCATTTGTCCATCAATATTATATTCAAGATATAAGTAAAAATAAACTTTCTGTAAACCAAGTTGTATGTTGTACTCAAGATAAC

CTCTCTTGTGTATTTGATTTTTGTATGAAAAAAACTAAAAATGGTAATCATACTTAATTATCAGTTATGGTAAATGGTATGAAGAGAAGAAG

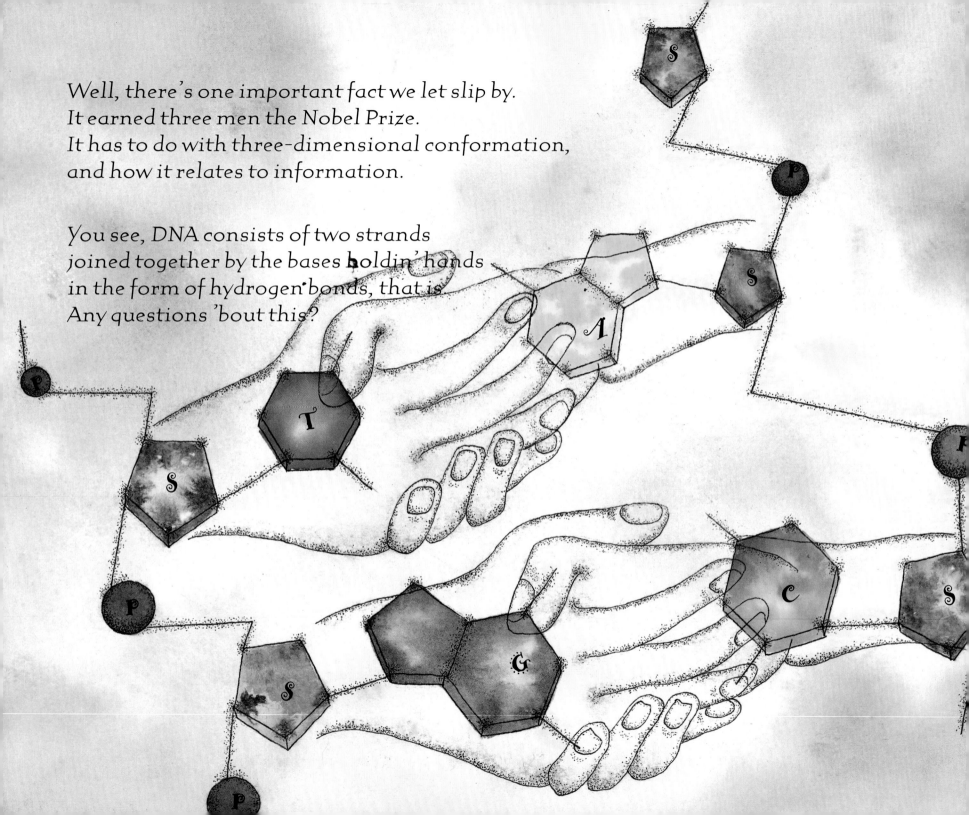

Well, there's one important fact we let slip by.
It earned three men the Nobel Prize.
It has to do with three-dimensional conformation,
and how it relates to information.

You see, DNA consists of two strands
joined together by the bases holdin' hands
in the form of hydrogen bonds, that is.
Any questions 'bout this?

The whole molecule forms a double helix—
A spiral staircase arrangement
with the nucleotide bases represented by the steps.
Hope this isn't too steep for you!

Maurice Wilkins

Francis Crick

James D. Watson

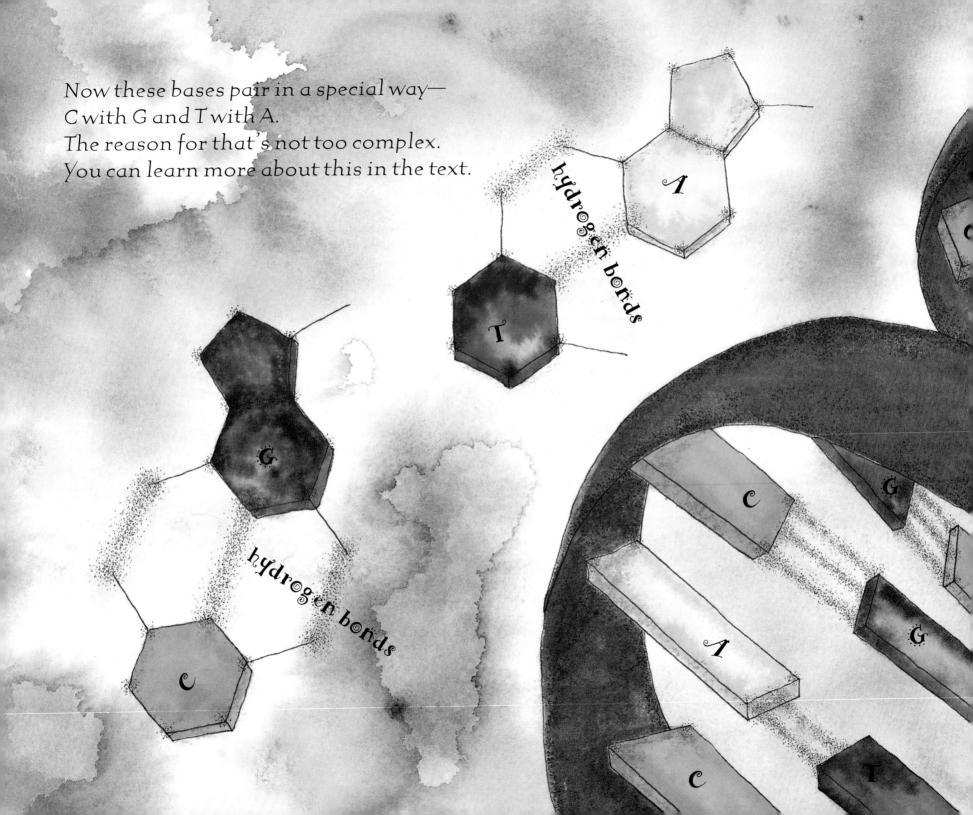

Now these bases pair in a special way—
C with G and T with A.
The reason for that's not too complex.
You can learn more about this in the text.

hydrogen bonds

hydrogen bonds

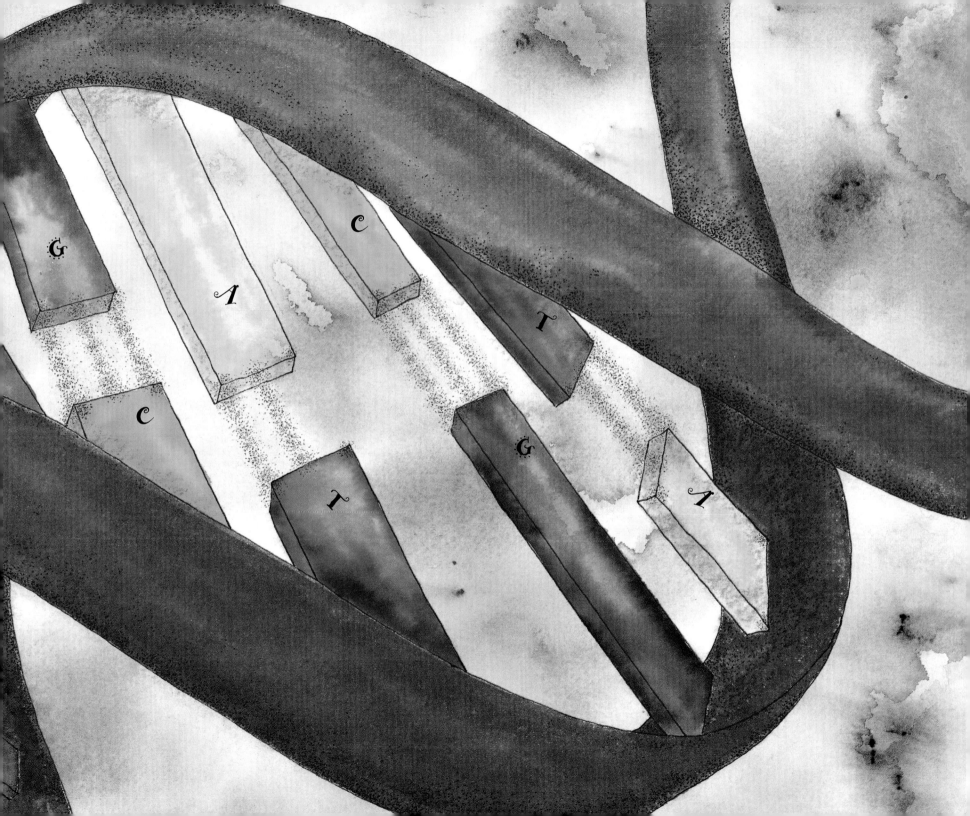

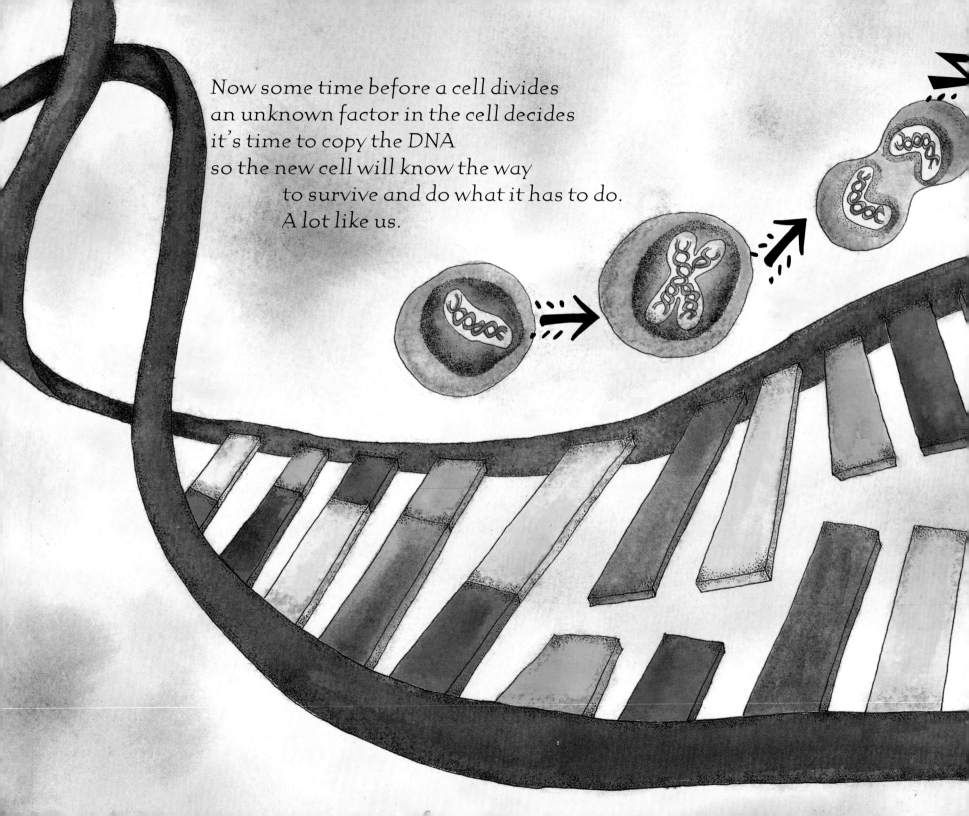

Now some time before a cell divides
an unknown factor in the cell decides
it's time to copy the DNA
so the new cell will know the way
to survive and do what it has to do.
A lot like us.

2 identical cells

G C

T A

G

C

A

T

Hence, the double helices unzip,
and nucleotides in the neighborhood slip
into place next to their proper mates.
I think you see what this creates—
Two daughter macromolecules
identical to the parent.

A

C T

G

A

T

G

C

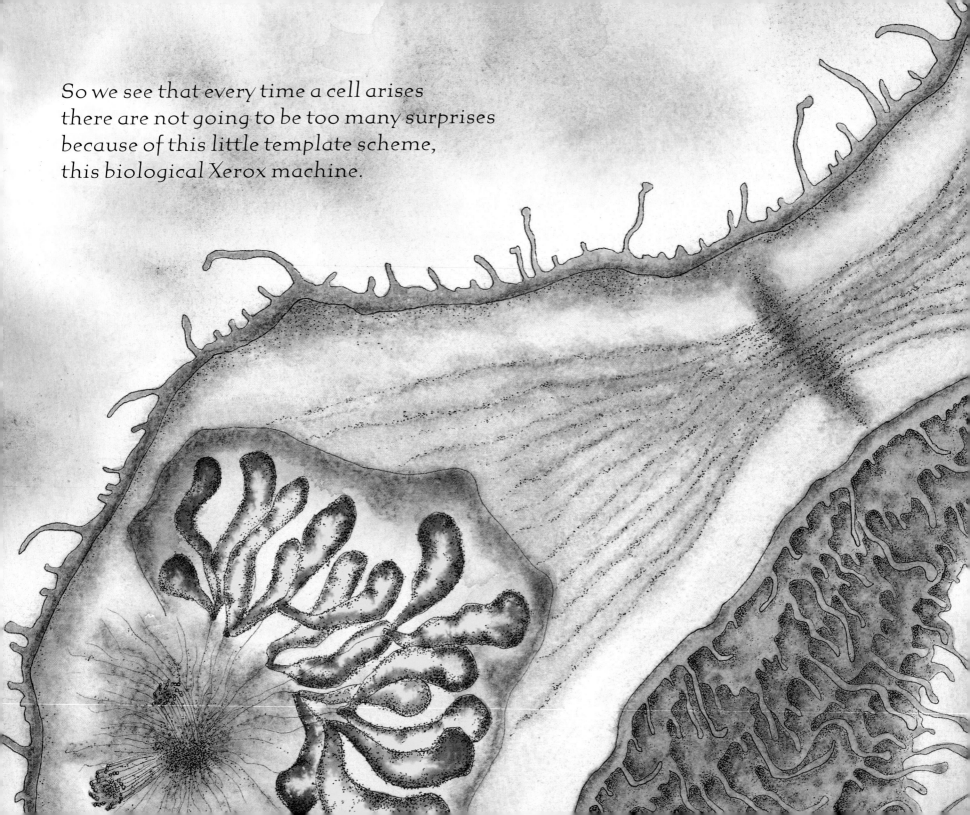

So we see that every time a cell arises
there are not going to be too many surprises
because of this little template scheme,
this biological Xerox machine.

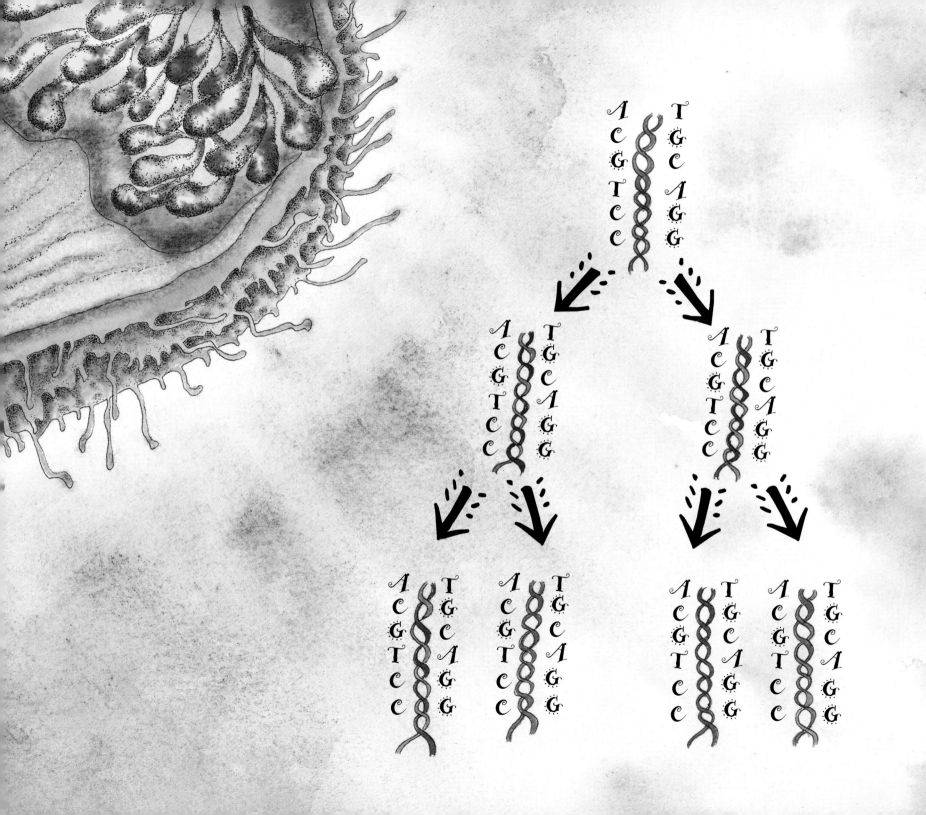

And when human parents make their contribution
to a baby-to-be's constitution—
Since their chromosomes are human, we presume,
it's no shock that the baby is, too.

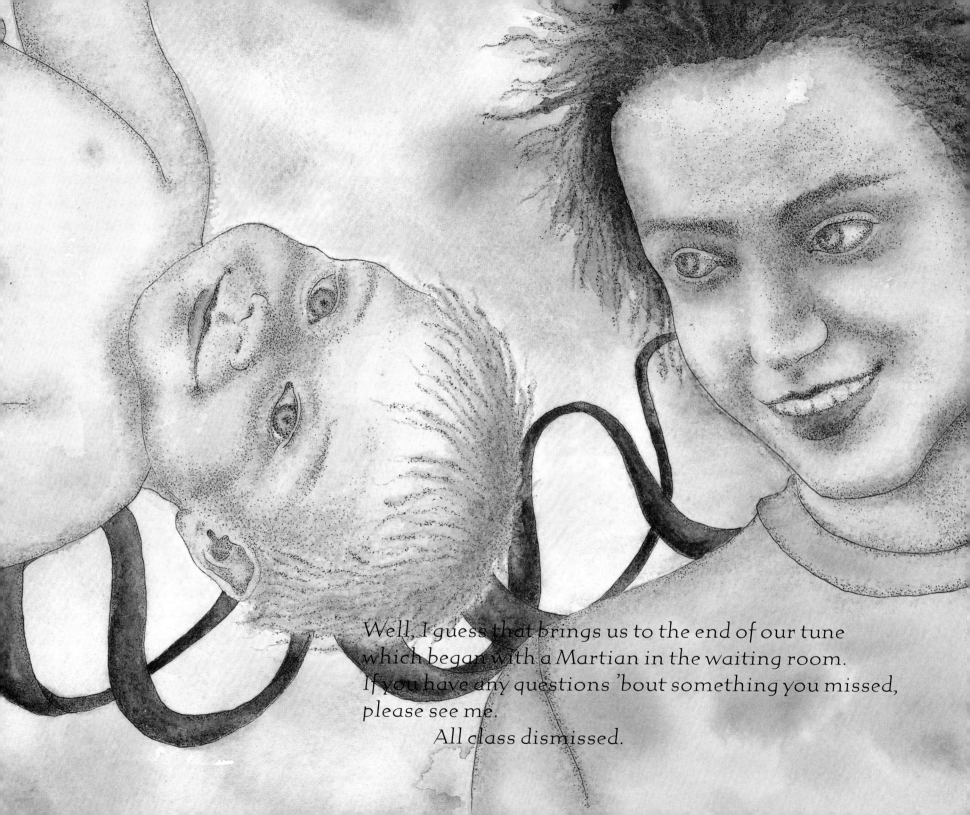

Well, I guess that brings us to the end of our tune
which began with a Martian in the waiting room.
If you have any questions 'bout something you missed,
please see me.
        All class dismissed.

A Double Helix Guide

Well, I once saw in an Addams cartoon
a Martian sittin' in a waiting room.
It was late at night in the maternity ward,
a nurse appeared as she opened the door—
  "Congratulations," she said "it's a baby."

Well, the point of this story, I'll tell you now,
did you ever sit down and think about how
it is that every time a baby's born
it's a baby—not a rabbit or an ear of corn?

Well, it just so happens that inside everyone
are tiny plans that tell how the job's to be done:
They're worth more to you than the family jewels.
They're stored in the form of molecules.
  Like everything else I guess.
  Only different, and kind of special.

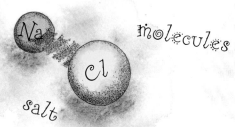

salt

molecules

water

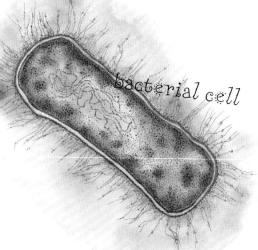

bacterial cell

intestinal cells

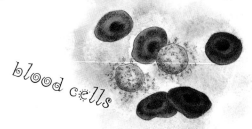

blood cells

It's a good thing to ask yourself or your friends or relatives or teachers how things get to be the way they are. They might not be able to give you an answer, but asking the question is the first step to finding an answer. One of the things I'm sure you've noticed is that humans always have human babies, not carrots or something else. Rabbits always produce baby rabbits, not hamsters or mice, and seeds from one kind of plant (like corn, for instance) always grow into the same kind of plant. Have you ever asked how this happens?

The reason is that inside every living creature there are plans that contain instructions for how that creature will grow. These instructions are also the reason why you, and any brothers and sisters you have, look like your parents. The instructions for making you are a combination of information from your father and information from your mother. The plans for making your brothers and sisters also came from your mother and father, but each one of them got slightly different information. That is why your brothers and sisters all look a bit like your parents and also a bit like you, but are not exactly alike (unless they are identical twins, who get exactly the same information from their mother and father).

A **molecule** is the smallest amount of a particular chemical. Molecules are combinations of atoms. There are over 100 different types of atoms, which can combine to make millions of different molecules. For example, a molecule of table salt contains two different atoms, one of sodium (abbreviated Na) and one of chlorine (abbreviated Cl). One molecule of water (H_2O) contains three atoms, two of hydrogen and one of oxygen.

The bodies of living creatures are made of **cells**. Some organisms such as bacteria, yeast, and algae consist simply of single cells, each living separately. Other larger and more complicated creatures such as humans are made of around one trillion cells—a million times one million cells. Cells are formed from other cells. One cell grows and then divides into two. Two cells become four, four become eight and so on. There are many different types of cell in a human body, each with different jobs to do (such as digesting food, producing hormones that make us grow, and covering our body with skin for protection). A single cell is so small that even a million of them cover just the tip of your finger. Though a cell is very small, it is highly organized and contains many different compartments, each designed for a different purpose. Of special importance to our story is one compartment called the **nucleus**. This is where the instructions are stored for making the different types of cells and for organizing them to form a whole living creature like you and me.

Inside the nucleus, these instructions are packaged in what are called **chromosomes**. If you look with a microscope at a cell that has been stained with a special dye, you can see many skinny, colored objects. These are the chromosomes. All organisms have them, but different organisms have different numbers of chromosomes. Rabbits have 42, corn has 20, dogs have 78, and humans have 46. Each type of chromosome has a different shape and pattern of staining with the colored dye. So, when we look at human chromosomes within a nucleus, we can say "That big one is chromosome 1, and that smaller one is chromosome 2" and so on. In most cells, there are two chromosomes of each kind: two copies of chromosome 1, two copies of chromosome 2, etc. In humans, that makes for a total of 23 pairs of chromosomes. We get one copy of chromosome 1 from our mother and one copy of chromosome 1 from our father; one copy of chromosome 2 from our mother and one copy of chromosome 2 from our father; etc. Each baby gets half of its chromosomes from its mother and half from its father. That is what we mean when we say that the instructions for making you are a combination of information from your mother and from your father. So, now we see that this information is packaged in the form of chromosomes. In 22 of the 23 pairs of human chromosomes, the two chromosomes of the pair look the same. In girls, the 23rd pair of chromosomes (called X chromosomes) looks the same. Boys have only one X chromosome (which they get from their mother). They also have a Y chromosome (which they get from their father).

Now, floatin' around in each one of your cells
is a nucleus, and you know what else—
tucked inside every one of these
are chromosomes that hold the keys
to the question we asked:
'bout how humans beget humans.

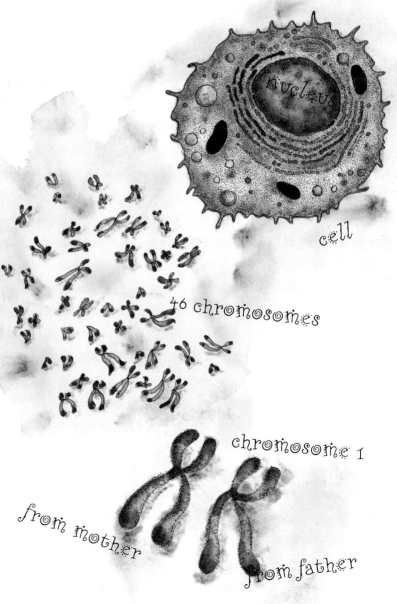

cell

46 chromosomes

chromosome 1

from mother

from father

skin cells

Now these threadlike chromosomes contain
a groovy little substance that we're gonna name;
it's a macromolecule, as they say.
It's mighty fine stuff—called DNA.
    That's "deoxyribonucleic acid"
    For those of you out there with expanding minds.

Now this DNA consists of a chain
of sugar and phosphate over and over again.
But just a minute's thought, and you'll see there's no hope
for such a simple molecule to contain all the dope
    to get the human show on the road
    and into high gear.

Well each of the sugars in the backbone
has a ring-shaped base 'tached to it all its own.
There's only four in a human being:
Adenine, guanine, cytosine, and thymine.
    They make up an alphabet of four letters—A, G, C, T.
    Wouldn't want to write a novel with four letters.
    Think I'll write a human being instead.

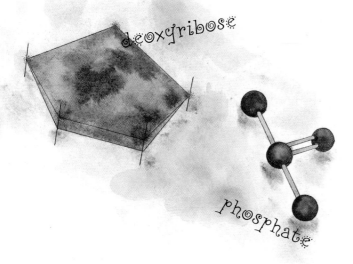

All chromosomes contain an enormously long molecule (a **macro-molecule**) called **deoxyribonucleic acid** or **DNA** for short. It's an acid and it's found in the nucleus, so it's called a nucleic acid. DNA is similar to another nucleic acid called ribonucleic acid (RNA for short), but it has one less oxygen atom. That's why it's called "**deoxy**ribonucleic acid." So, you see that even though its name is long, when it is broken down, it's easy to understand what the name "deoxyribonucleic acid" means. We are going to use the same plan to understand the DNA molecule itself: break it into small pieces so that its overall structure can be understood.

DNA contains many molecules of **sugar**. The type of sugar found in DNA is different from the type of sugar that you eat, but it is related to it. Sugars are molecules that contain a particular arrangement of carbon atoms, hydrogen atoms, and oxygen atoms. The sugar in DNA is called deoxyribose. A molecule of deoxyribose consists of 5 atoms—4 carbons and 1 oxygen—that join together to form a ring. Other atoms are joined to this ring. The sugar molecules are linked together by other molecules called **phosphates**. Each one of these phosphates contains 1 atom of phosphorus and 3 atoms of oxygen. The chain of sugar and phosphate molecules is incredibly long. Each human chromosome contains a single molecule of DNA with around 280 million sugars and 280 million phosphates!

This chain of sugars and phosphates is the **backbone** of the DNA molecule. What makes DNA special is that each sugar molecule is attached to another kind of molecule called a base. There are four different kinds of bases in DNA: **adenine, guanine, cytosine,** and **thymine,** known by their initials A, G, C, and T. Each base consists of a different combination of atoms of carbon, hydrogen, oxygen, and nitrogen. A and G are similar to each other because they are both made of two rings that are joined together. But A and G are also different from each other because they have different atoms attached to one of their rings. So, molecules of A and G differ from each other on one side only. C and T are both made of a single ring, but they are also different from each other because the atoms attached to this ring are

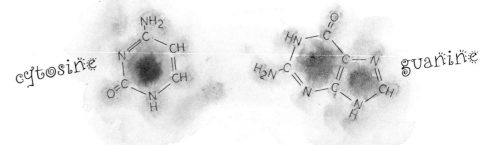

different. So, molecules of C and T also differ from each other on one side only. These different sides of A, G, C, and T play incredibly important roles in answering our question.

For their part in figuring out the shape of DNA, three men were awarded the **Nobel Prize**. This prize recognizes achievements in different areas (in chemistry, physics, literature, economics, peace, and physiology or medicine) and is named after Alfred Nobel, a Swedish inventor who left money in his will for these prizes. The work done by James Watson, Francis Crick, and Maurice Wilkins showed the shape of a DNA molecule in three dimensions—what is called its **three-dimensional conformation**. When they discovered this shape, Watson and Crick immediately understood the answer to the question "Why do humans produce a human baby and not a rabbit or an ear of corn?".

Imagine that you're a detective and have just found the clue that gives you the answer to a mystery. All of a sudden, everything falls into place. This is what happened when Watson and Crick found that DNA contains not just one but two chains of sugars and phosphates. Let's first look at the structure of these two chains and how they are held together. Then we can see how this **two**-chain structure of DNA explains the "mystery" of how humans produce human babies and not rabbits or ears of corn.

The two chains of DNA are joined together because each base on one chain shares hydrogen atoms with one of the bases on the opposite chain. This sharing of hydrogen atoms is called a **hydrogen bond**. G and C form three hydrogen bonds, which is like holding hands with three fingers. A and T form two hydrogen bonds, which is like holding hands with two fingers. Together, the two chains make a shape like a staircase. In this staircase, each step is a little higher than the last and a little to the left. So the staircase goes upward in a spiral, a shape that is called a helix. Because DNA has two backbone chains like this, it is called a **double helix**. A **nucleotide** is a term that describes a base attached to a sugar attached to a phosphate molecule.

Well, there's one important fact we let slip by.
It earned three men the Nobel Prize.
It has to do with three-dimensional conformation,
and how it relates to information.

You see, DNA consists of two strands
joined together by the bases holdin' hands
in the form of hydrogen bonds, that is.
Any questions 'bout this?
    The whole molecule forms a double helix—
    A spiral staircase arrangement
    with the nucleotide bases represented by the steps.
    Hope all this isn't too steep for you!

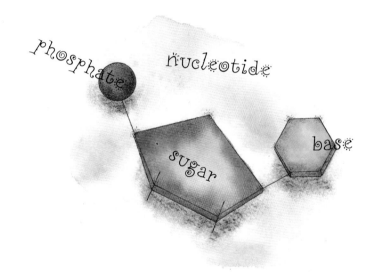

Now these bases pair in a special way—
C with G and T with A.
The reason for that's not too complex.
You can learn more about this in the text.

Now some time before a cell divides
an unknown factor in the cell decides
it's time to copy the DNA
so the new cell will know the way
    to survive and do what it has to do.
    A lot like us.

Hence, the double helices unzip,
and nucleotides in the neighborhood slip
into place next to their proper mates.
I think you see what this creates—
    Two daughter macromolecules
    identical to the parent.

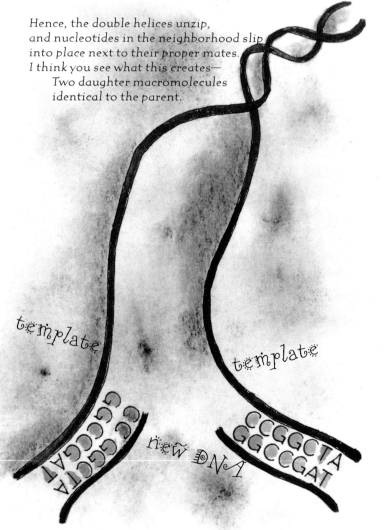

The joining of two of these bases by hydrogen bonds is called **pairing**. C can pair only with G. T can pair only with A. Pairs are made between a base with one ring (C or T) and a base with two rings (G and A), so that space between the two chains of DNA is always the same.

The nucleus of a human cell contains 46 chromosomes—so, it's got 46 long DNA molecules. New cells are made by the division of existing cells. When the time comes for this to happen, the cell gets bigger (it doubles in size), and then divides into two cells of equal size. In order for these new cells to work properly, each one has to end up with the right number of molecules of DNA—46 of them, each packaged into a chromosome. So, a cell that's preparing to divide first makes a copy of each of its 46 chromosomes. Then it gives each new cell a copy of each of these chromosomes. Copying a chromosome means copying a molecule of DNA.

How is a molecule of DNA copied? This is where the structure of DNA with its two chains becomes so important. First, the hydrogen bonds break between the bases in the two chains of the double helix. This allows the bases in each chain to make new hydrogen bonds with nucleotides floating around inside the nucleus that haven't yet been joined together. Of course, the pairing rules still apply, C with G and T with A. Each base in an existing DNA chain pairs with a previously unattached partner, and these newly attached bases join together to form a new chain that builds up alongside the original one.

In this way, one DNA molecule with its two nucleotide chains is copied to make two new DNA molecules. In each new molecule, only one of the two chains is newly made. The sequence of bases in this new chain is the same as the sequence in the original chain's previous partner. So, each chain of DNA is a **template** for copying the sequence of its partner chain. Before cells can divide, the DNA in each of our chromosomes is copied so that each new cell can receive one copy of each chromosome. Because the DNA molecule of a chromosome is so long, it first has to be scrunched up into a much more compact form. In this form, a chromosome becomes visible in a microscope and looks like the chromosomes that you can see throughout the book. When the cell divides into two, each new cell gets one of each kind of chromosome. And because each new cell has the same chromosomes, they both behave just like the cells they came from, because it's the DNA that tells them what kind of cells they are.

The first cell of a baby (called a zygote) is formed when a cell from the father (a sperm cell) joins with a cell from the mother (an egg cell). A sperm cell has only 23 chromosomes (from the father), and an egg cell has only 23 chromosomes (from the mother). When the egg and sperm cells combine, the zygote ends up with two sets of chromosomes—46 in all. Because these chromosomes contain DNA molecules from human parents, the baby that grows from the zygote is—no surprise—a human being!

The cartoon about the Martian in the waiting room makes us appreciate something that many people take for granted—how remarkable it is that when a baby's born, it's a baby, not a rabbit or an ear of corn. We have seen in this song and in this book that the answer to this fascinating question—this "mystery" that has puzzled people for thousands of years—is simple. The answer lies in understanding the structure of a molecule and how this molecule is used to make copies of itself by a simple but elegant procedure.

It is amazing to realize that it's the same molecule—DNA—in people and in all other living things that contains the information that makes them all different and special. How this DNA information provides the instructions to build a tree or a turkey, a polar bear or a person, is another story. Only some of that story is known right now. Perhaps you will help figure out the rest.

So we see that every time a cell arises
there are not going to be too many surprises
because of this little template scheme,
this biological Xerox machine.

And when human parents make their contribution
to a baby-to-be's constitution—
Since their chromosomes are human, we presume,
it's no shock that the baby is, too.

Well I guess that brings us to the end of our tune
which began with a Martian in the waiting room.
If you have any questions 'bout something you missed,
please see me.
   All class dismissed.

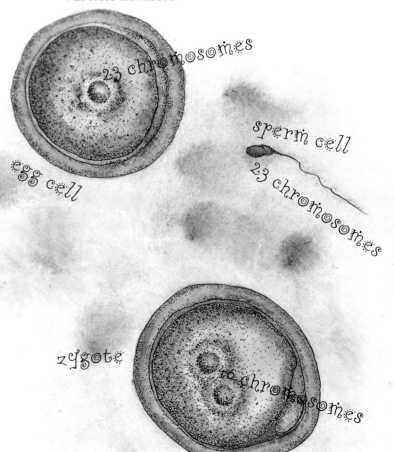

23 chromosomes

egg cell

sperm cell

23 chromosomes

zygote

46 chromosomes

More Information About DNA

Arnold, Caroline. 1986. Genetics: From Mendel to Gene Splicing. Watts, New York.

Asimov, Isaac. 1983. How Did We Find Out About Genes. Illustrated by David Wool. Walker and Company, New York.

Asimov, Isaac. 1985. How Did We Find Out About DNA. Illustrated by David Wool. Walker and Company, New York.

Balkwill, Fran. 1992. DNA Is Here to Stay. Illustrated by Mic Rolph. HarperCollins Publishers, UK.

Balkwill, Fran. 1993. Amazing Schemes Within Your Genes. Illustrated by Mic Rolph. HarperCollins Publishers, UK.

Bornstein, Sandy. 1989. What Makes You What You Are: A First Look at Genetics. Simon & Schuster, New York.

Burnie, David. 1991. How Nature Works. Reader's Digest Association, Inc., Pleasantville, New York, Toronto/Dorling Kindersley Limited, London.

Edelson, Edward. 1991. Genetics and Heredity. Chelsea House

Facklam, Margery and Howard. 1979. From Cell to Clone. Harcourt Brace Jovanovich, New York, London.

Fradin, Dennis. 1987. Heredity. Childrens Press, Chicago.

Gonick, Larry, and Wheelis, Mark. 1991. The Cartoon Guide to Genetics. HarperCollins, New York.

Palent, Dorothy Henshaw. 1989. Grandfather's Nose: Why We Look Alike or Different. Illustrated by Diane Palmisciano. Franklin Watts, New York.

Showers, Paul, 1978. Me and My Family Tree. Thomas Y. Crowell, New York.

Van Loon, Borin. 1991. DNA—The Marvelous Molecule. Tarquin, UK.

Wilcox, Frank. 1988. DNA: The Thread of Life. Lerner Publishers.

Acknowledgements

The authors would like to thank Reida and Irwin Herskowitz for comments on the Double Helix Guide, and Lynn Rubin and John Inglis for encouragement and suggestions. We also thank the Cold Spring Harbor Laboratory Archives for reference photographs of Maurice Wilkins, Francis Crick, and James Watson.

The type for this book was set in Emigre Remedy and Emigre Quartet by The Sarabande Press, New York, New York. The illustrations were done with a combination of inks, dyes, watercolors, pastels, and colored pencils on Arches paper.